This Book Belongs to

_____

**FORMULAE
AT THE END OF
THE BOOK**

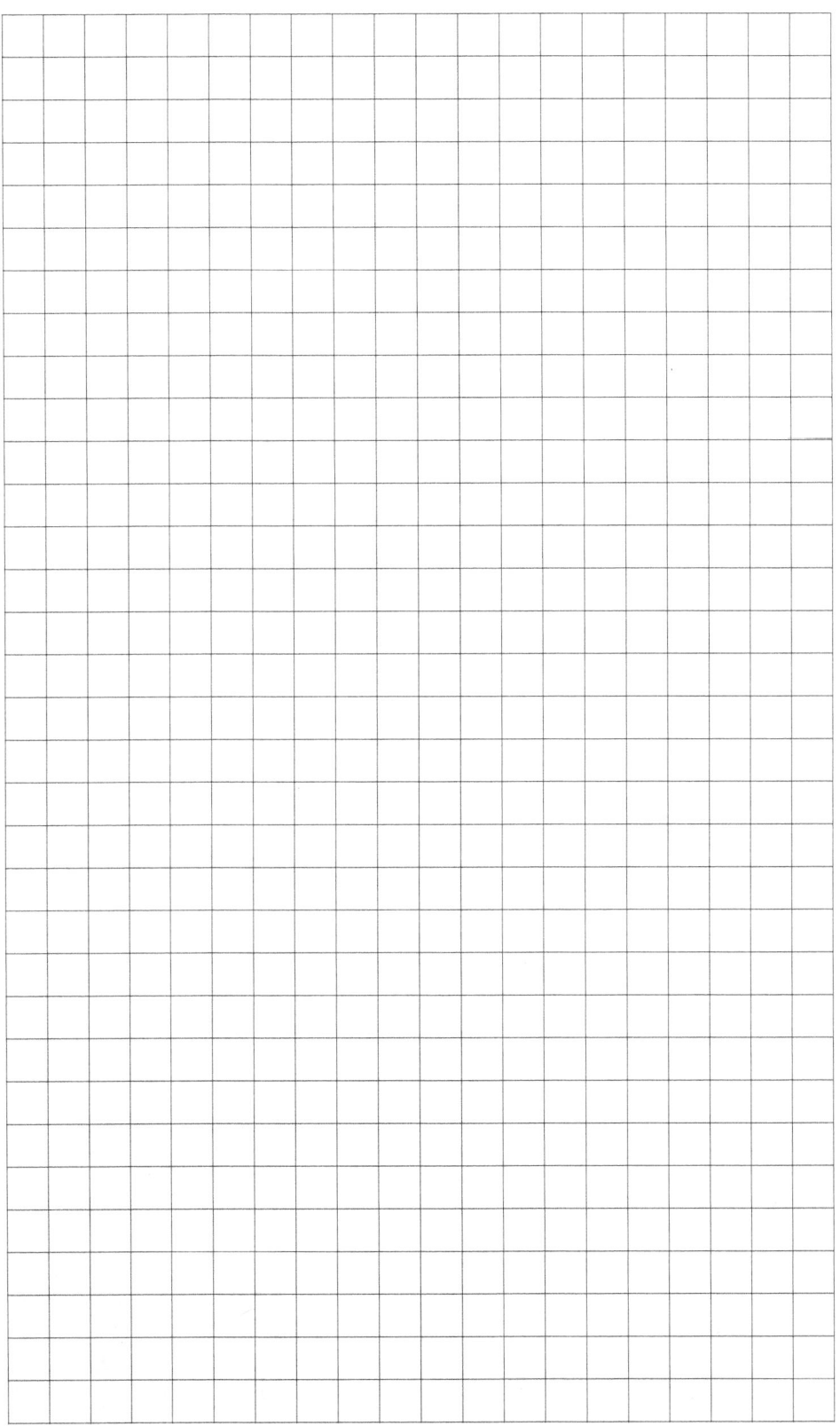

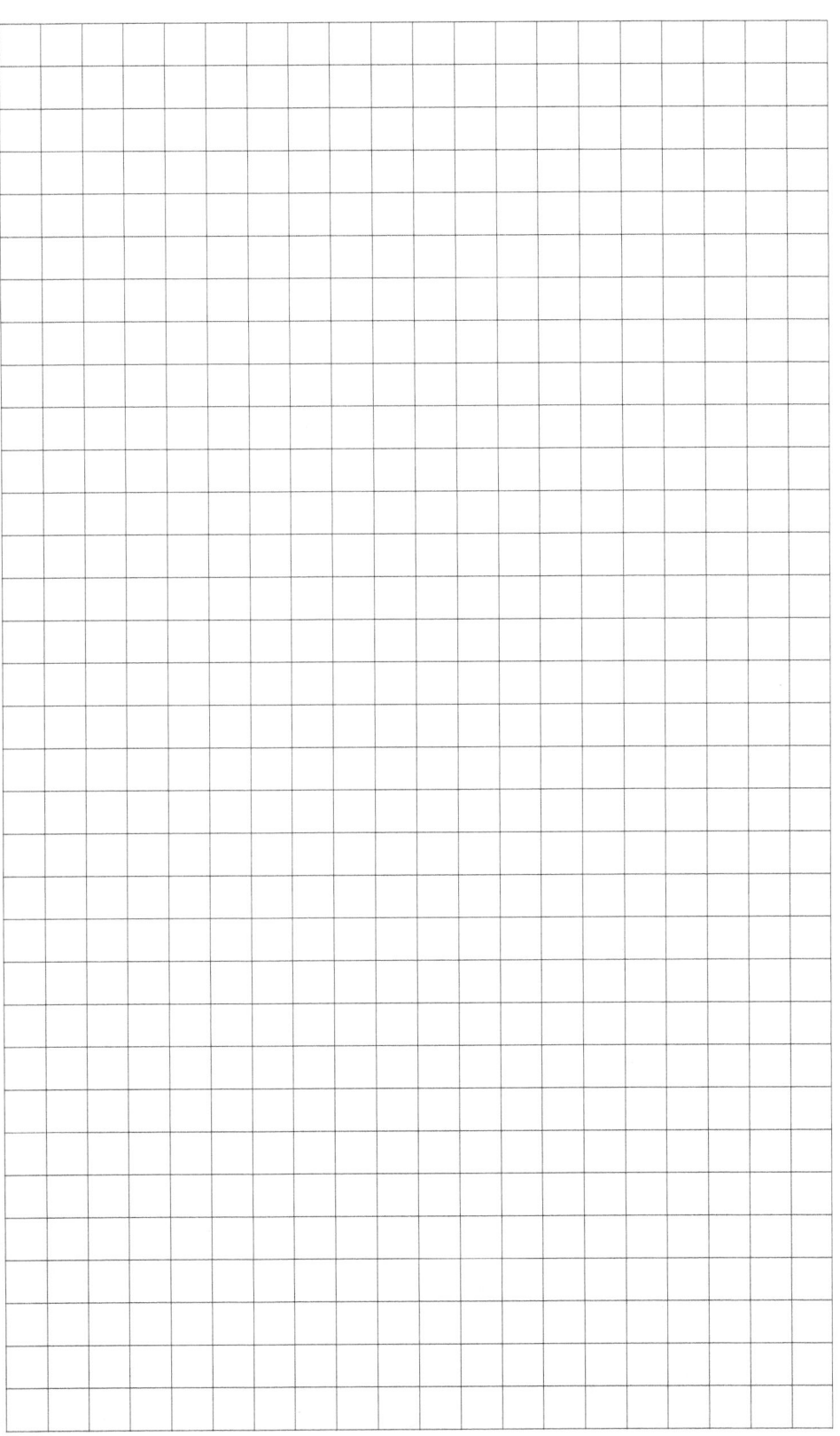

SI units

SI Derived Units

Unit	Symbol	In SI Units	Quantity
Mechanics			
Pascal	Pa	kg·m⁻¹·s⁻²	Pressure, Stress
Joule	J	kg·m²·s⁻²	Energy, Work, Heat
Watt	W	kg·m²·s⁻³	Power
Newton	N	kg·m·s⁻²	Force, Weight
Electromagnetism			
Tesla	T	kg·s⁻²·A⁻¹	Magnetic Field
Henry	H	kg·m²·s⁻²·A⁻²	Inductance
Coulomb	C	A·s	Electric Charge
Volt	V	kg·m²·s⁻³·A⁻¹	Voltage
Farad	F	kg⁻¹·m⁻²·s⁴·A²	Electric Capacitance
Siemens	S	kg⁻¹·m⁻²·s³·A²	Electrical Conductance
Weber	Wb	kg·m²·s⁻²·A⁻¹	Magnetic Flux
Ohm	Ω	kg·m²·s⁻³·A⁻³	Electric Resistance
Optics			
Lux	lx	cd·sr·m⁻²	Illuminance
Lumen	lm	cd·sr	Luminous Flux
Radioactivity			
Becquerel	Bq	s⁻¹	Radioactivity
Gray	Gy	m²·s⁻¹	Absorbed Dose
Sievert	Sv	m²·s⁻¹	Equivalent Dose
Other			
Hertz	Hz	s⁻¹	Frequency
Katal	kat	mol·s⁻¹	Catalytic Activity

SI base units list

Unit	Symbol	Quantity
Meter (metre)	m	Length
Kilogram	kg	Mass
Second	s	Time
Ampere	A	Electric current
Kelvin	K	Thermodynamic temperature
Mole	mol	Amount of substance
Candela	cd	Luminous intensity

SI Supplementary Units

Unit	Symbol	Quantity
Radian	rad	Plane angle (2D angle)
Steradian	sr	Solid angle (3D angle)

Area

square	$A = l^2$	l : length of side
Rectangle	$A = w \times h$	w : width h : height
Triangle	$A = \dfrac{b \times h}{2}$	b : base h : height
Rhombus	$A = \dfrac{D \times d}{2}$	D : large diagonal d : small diagonal
Trapezoid	$A = \dfrac{B + b}{2} \times h$	B : large side b : small side h : height
Regular polygon	$A = \dfrac{P}{2} \times a$	P : perimeter a : apothem
Circle	$A = \pi r^2$ $P = 2\pi r$	r : radius P : perimeter
Cone (lateral surface)	$A = \pi r \times s$	r : radius s : slant height
Sphere (surface area)	$A = 4\pi r^2$	

Statistics

Summation Rules and Properties	$\sum_{i=p}^{n} \lambda = (n-p+1)\lambda$	
	$\sum_{i=1}^{n} \lambda x_i = \lambda \sum_{i=1}^{n} x_i$	
	$\sum_{i=1}^{n} (x_i + y_i) = \sum_{i=1}^{n} x_i + \sum_{i=1}^{n} y_i$	
	$\sum_{i=1}^{n} x_i = \sum_{i=1}^{p} x_i + \sum_{i=p+1}^{n} x_i$	
Used Symbols	Statistical sample	$x = (x_1, x_2, x_3, \ldots, x_n)$
	Sample size	N
	Absolute Frequency	n_i
	Relative Frequency	$f_i = \dfrac{n_i}{N}$
	Cumulative (Absolute) Frequency	N_i
	Cumulative Relative Frequency	F_i
Sample Mean	Ungrouped Data	$\bar{x} = \dfrac{\sum_{i=1}^{k} x_i}{N}$
	Grouped Data	$\bar{x} = \dfrac{\sum_{i=1}^{k} n_i x_i}{N}$
		$\bar{x} = \sum_{i=1}^{k} f_i x_i$
Median	If N is odd	$Me = x_k,\ k = \dfrac{N+1}{2}$
	If N is even	$Me = \dfrac{x_k + x_{k+1}}{2},\ k = \dfrac{N}{2}$
Sum of Deviations from the Mean	$\sum_{i=1}^{k} d_i = \sum_{i=1}^{k} (x_i - \bar{x}) = 0$	
Sum of Squared Deviations from the Mean	Ungrouped Data	$SS_x = \sum_{i=1}^{k} (x_i - \bar{x})^2$
		$SS_x = \sum_{i=1}^{k} x_i^2 - k\bar{x}^2$
	Grouped Data	$SS_x = \sum_{i=1}^{k} (x_i - \bar{x})^2 n_i$
Sample Variance	$S_x^2 = \dfrac{SS_x}{N-1}$	
Sample Standard Deviation	$S_x = \sqrt{\dfrac{SS_x}{N-1}}$	

Exponents

Product	$a^m \times a^n = a^{m+n}$	ex: $3^5 \times 3^2 = 3^{5+2} = 3^7$
	$a^m \times b^m = (a \times b)^m$	ex: $3^5 \times 2^5 = (3 \times 2)^5 = 6^5$
Quotient	$a^m \div a^n = a^{m-n}$	ex: $3^7 \div 3^2 = 3^{7-2} = 3^5$
	$a^m \div b^m = (a \div b)^m$	ex: $6^5 \div 2^5 = (6 \div 2)^5 = 3^5$
		ex: $5^3 \div 2^3 = (\frac{5}{2})^3$
Power of Power	$(a^m)^p = a^{m \times p}$	ex: $(5^2)^3 = 5^{2 \times 3} = 5^6$
Zero Exponents	$a^0 = 1$	ex: $8^0 = 1$
Negative Exponents	$a^{-n} = (\frac{1}{a})^n$	ex: $3^{-2} = (\frac{1}{3})^2$
		ex: $(\frac{2}{3})^{-4} = (\frac{3}{2})^4$
Fractional Exponents	$a^{\frac{p}{q}} = \sqrt[q]{a^p}$	ex: $24^{\frac{1}{3}} = \sqrt[3]{24}$

Directly Proportional		$y=kx$	$k=\frac{y}{x}$
			k: Constant of Proportionality
Inversely Proportional		$y=\frac{k}{x}$	$k=yx$

	Quadratic formula		$x=\dfrac{-b\pm\sqrt{b^2-4ac}}{2a}$
$ax^2+bx+c=0$	Concavity		Concave up: $a>0$
			Concave down: $a<0$
	Discriminant		$\Delta=b^2-4ac$
	Vertex of the parabola		$V\left(-\dfrac{b}{2a},-\dfrac{\Delta}{4a}\right)$
$y=a(x-h)^2+k$	Concavity		Concave up: $a>0$
			Concave down: $a<0$
	Vertex of the parabola		$V(h,k)$

Zero-product property	$A\times B=0 \Leftrightarrow A=0 \vee B=0$	ex: $(x+2)\times(x-1)=0 \Leftrightarrow x+2=0 \vee x-1=0 \Leftrightarrow x=-2 \vee x=1$
Difference of two squares	$(a-b)(a+b)=a^2-b^2$	ex: $(x-2)(x+2)=x^2-2^2=x^2-4$
Perfect square trinomial	$(a+b)^2=a^2+2ab+b^2$	ex: $(2x+3)^2=(2x)^2+2\cdot 2x\cdot 3+3^2=4x^2+12x+9$
Binomial theorem	$(x+y)^n=\sum_{k=0}^{n} {}_nC_k\, x^{n-k}\, y^k$	

LOG

Definition	$\log_a b = x \Leftrightarrow b = a^x$	ex: $3^x = 15 \Leftrightarrow x = \log_3 15$
	$\log_a 1 = 0$	ex: $\log_3 1 = 0$
	$\log_a a = 1$	ex: $\log 10 = 1$
	$\log_a a^b = b$	ex: $\ln e^2 = 2$
Product	$\log_a(u \times v) = \log_a u + \log_a v$	ex: $\log_6 10 + \log_6 2 = \log_6(10 \times 2) = \log_6 20$
Quotient	$\log_a(u/v) = \log_a u - \log_a v$	ex: $\log_4 9 - \log_4 3 = \log_4(9/3) = \log_4 3$
Exponential	$\log_a u^v = v \times \log_a u$	ex: $\log_4 36 = \log_4 6^2 = 2 \times \log_4 6$
Change of Base	$\log_a u = \dfrac{\log_b u}{\log_b a}$	ex: $\log_4 5 \times \log_5 6 = \log_4 5 \times \dfrac{\log_4 6}{\log_4 5} = \log_4 6$

	Volume	
Cube	$V=s^3$	s: side
Parallelepiped	$V = l \times w \times h$	l: length w: width h: height
Regular prism	$V = b \times h$	b: base h: height
Cylinder	$V = \pi r^2 \times h$	r: radius h: height
Cone (or pyramid)	$V = \frac{1}{3} b \times h$	b: base h: height
Sphere	$V = \frac{4}{3} \pi r^3$	r: radius

Trigonometry

Trigonometry Ratios

$\sin\alpha = \frac{opp}{hip}$ opp.: opposite, hip.: hypotenuse

$\cos\alpha = \frac{adj}{hip}$ adj.: adjacent, hip.: hypotenuse

$\tan\alpha = \frac{opp}{adj}$ opp.: opposite, adj.: adjacent

Fundamental Identities

$\sin^2\alpha + \cos^2\alpha = 1$ $\tan\alpha = \frac{\sin\alpha}{\cos\alpha}$ $\tan^2\alpha + 1 = \frac{1}{\cos^2\alpha}$

Law of Sines (aka sine rule): $\frac{\sin A}{a} = \frac{\sin B}{b} = \frac{\sin C}{c}$

Law of Cosines (aka cosine rule): $a^2 = b^2 + c^2 - 2bc\cos A$

Heron's formula: $A = \sqrt{s(s-a)(s-b)(s-c)}$, $s = \frac{a+b+c}{2}$

(Diagram: right triangle with hypotenuse, opposite, adjacent, angle α)

Exact Values

$\sin(\pi/6) = 1/2$	$\cos(\pi/6) = \sqrt{3}/2$	$\tan(\pi/6) = \sqrt{3}/3$
$\sin(\pi/4) = \sqrt{2}/2$	$\cos(\pi/4) = \sqrt{2}/2$	$\tan(\pi/4) = 1$
$\sin(\pi/3) = \sqrt{3}/2$	$\cos(\pi/3) = 1/2$	$\tan(\pi/3) = \sqrt{3}$

Angle Relationships

$\sin(-\alpha) = -\sin\alpha$	$\cos(-\alpha) = \cos\alpha$	$\tan(-\alpha) = -\tan\alpha$
$\sin(\pi-\alpha) = \sin\alpha$	$\cos(\pi-\alpha) = -\cos\alpha$	$\tan(\pi-\alpha) = -\tan\alpha$
$\sin(\pi+\alpha) = -\sin\alpha$	$\cos(\pi+\alpha) = -\cos\alpha$	$\tan(\pi+\alpha) = \tan\alpha$
$\sin(\pi/2-\alpha) = \cos\alpha$	$\cos(\pi/2-\alpha) = \sin\alpha$	$\tan(\pi/2-\alpha) = 1/\tan\alpha$
$\sin(\pi/2+\alpha) = \cos\alpha$	$\cos(\pi/2+\alpha) = -\sin\alpha$	$\tan(\pi/2+\alpha) = -1/\tan\alpha$
$\sin(3\pi/2-\alpha) = -\cos\alpha$	$\cos(3\pi/2-\alpha) = -\sin\alpha$	$\tan(3\pi/2-\alpha) = 1/\tan\alpha$
$\sin(3\pi/2+\alpha) = -\cos\alpha$	$\cos(3\pi/2+\alpha) = \sin\alpha$	$\tan(3\pi/2+\alpha) = -1/\tan\alpha$

Trigonometric Equations

$\sin x = \sin\alpha \Leftrightarrow x = \alpha + 2k\pi \lor x = \pi - \alpha + 2k\pi, \; k \in \mathbb{Z}$

$\cos x = \cos\alpha \Leftrightarrow x = \alpha + 2k\pi \lor x = -\alpha + 2k\pi, \; k \in \mathbb{Z}$

$\tan x = \tan\alpha \Leftrightarrow x = \alpha + k\pi, \; k \in \mathbb{Z}$

Sum Formulas

$\sin(a+b) = \sin a \cos b + \sin b \cos a$

$\cos(a+b) = \cos a \cos b - \sin a \sin b$

$\tan(a+b) = \dfrac{\tan a + \tan b}{1 - \tan a \tan b}$

Difference Formulas

$\sin(a-b) = \sin a \cos b - \sin b \cos a$

$\cos(a-b) = \cos a \cos b + \sin a \sin b$

$\tan(a-b) = \dfrac{\tan a - \tan b}{1 + \tan a \tan b}$

Double Angle Formulas

$\sin(2a) = 2 \sin a \cos a$

$\cos(2a) = \cos^2 a - \sin^2 a$

$\tan(2a) = \dfrac{2 \tan a}{1 - \tan^2 a}$

www.ingramcontent.com/pod-product-compliance
Lightning Source LLC
Chambersburg PA
CBHW060836220526
45466CB00003B/1131